SCORPIONS

by Matt Lilley

Cody Koala
An Imprint of Pop!
popbooksonline.com

abdobooks.com
Published by Pop!, a division of ABDO, PO Box 398166, Minneapolis, Minnesota 55419. Copyright © 2022 by Abdo Consulting Group, Inc. International copyrights reserved in all countries. No part of this book may be reproduced in any form without written permission from the publisher. Cody Koala™ is a trademark and logo of Pop.

Printed in the United States of America, North Mankato, Minnesota.

052021
092021
THIS BOOK CONTAINS RECYCLED MATERIALS

Cover Photo: Shutterstock Images
Interior Photos: Shutterstock Images, 1, 5 (top), 5 (bottom left), 5 (bottom right), 6, 10, 11, 15, 19 (top), 19 (bottom left), 19 (bottom right); George Gerster/Science Source, 9; iStockphoto, 12–13, 16–17, 20

Editor: Aubrey Zalewski
Series Designers: Laura Graphenteen and Colleen McLaren

Library of Congress Control Number: 2020948831
Publisher's Cataloging-in-Publication Data
Names: Lilley, Matt, author.
Title: Scorpions / by Matt Lilley
Description: Minneapolis, Minnesota : Pop!, 2022 | Series: Desert animals | Includes online resources and index.
Identifiers: ISBN 9781532169748 (lib. bdg.) | ISBN 9781098240677 (ebook)
Subjects: LCSH: Scorpions--Juvenile literature. | Scorpions--Behavior--Juvenile literature. | Arachnids--Juvenile literature. | Desert animals--Juvenile literature.
Classification: DDC 591.754--dc23

Hello! My name is

Cody Koala

Pop open this book and you'll find QR codes like this one, loaded with information, so you can learn even more!

Scan this code* and others like it while you read, or visit the website below to make this book pop.

popbooksonline.com/scorpions

*Scanning QR codes requires a web-enabled smart device with a QR code reader app and a camera.

Table of Contents

Chapter 1
Desert Scorpions 4

Chapter 2
Hunting at Night. 8

Chapter 3
Life in the Desert 14

Chapter 4
Scorpion Babies. 18

Making Connections 22
Glossary. 23
Index 24
Online Resources 24

Chapter 1

Desert Scorpions

Scorpions have eight legs. They have two **pincers** and a stinger. **Exoskeletons** surround scorpions' bodies.

Watch a video here!

Scorpions live all over the world. Most kinds of scorpions live in the desert. Desert scorpions are often yellow or light brown. They match the color of the sand.

Chapter 2

Hunting at Night

There is not much food in the desert. Scorpions eat what they can find. Common **prey** are insects, lizards, and mice.

Scorpions hunt at night.

They have up to 12 eyes.

They can see things move in the dark.

Scorpions can go one year without eating.

Scorpions also have hairs that help them hunt. The hairs sense movement through the air or ground.

Scorpions wait for their prey. When it gets close, they attack. Scorpions grab and crush with their **pincers**. They have stingers with **venom**. They use their stingers so large prey cannot escape.

Chapter 3

Life in the Desert

Scorpions need to hide from the sun and other animals. They stay where it is cool. Scorpions rest under rocks or in **burrows**.

Many scorpions dig their own burrows.

Learn more here!

The desert is dry. Scorpions get the water they need from their food.

They save energy by not moving much.

> Scorpion **exoskeletons** have a layer of wax. It keeps water in.

Chapter 4

Scorpion Babies

Scorpions can give birth to more than 100 live babies. The babies crawl onto their mother's back. They stay safe up there for approximately 50 days.

Complete an activity here!

The babies have their first **molt**. Then they go off on their own. They grow with each molt. Scorpions molt four to nine times. They can live between five and 25 years.

Making Connections

Text-to-Self

Have you ever seen a scorpion in real life? If so, what did you do? If not, what do you think you would do if you saw one?

Text-to-Text

Have you read books about other animals that live in the desert? How are they similar to or different from scorpions?

Text-to-World

Scorpions have stingers. What other animals have stingers? What do they use them for?

Glossary

burrow – a hole that an animal digs in the ground for shelter.

exoskeleton – a hard outer layer that covers an animal's body.

molt – to shed an outer layer.

pincer – a front claw used to grab and hold things.

prey – an animal that is hunted by other animals.

venom – poison from an animal bite or sting.

Index

babies, 18, 21

burrows, 14, 15

exoskeletons, 4, 17

hairs, 11

molting, 21

pincers, 4, 12–13

prey, 8, 12

stingers, 4, 12–13

Online Resources

popbooksonline.com

Thanks for reading this Cody Koala book!

Scan this code* and others like it in this book, or visit the website below to make this book pop!

popbooksonline.com/scorpions

*Scanning QR codes requires a web-enabled smart device with a QR code reader app and a camera.

MAY 31 2022